Y0-AGK-190

CONTENTS

NOTICE OF DISCLAIMER

The author of this book is not a professional engineer nor has he had formal training in the design or operation of the project. The author is an amateur but has been successful in building and operating the project discussed herein.

The methods that he used and describes are presented merely as guidelines for other amateurs in developing such a project. The project can be dangerous, and dangers have been pointed out wherever possible. Since the author is not a professional in this field, there may well be other dangers involved in the building and operation of the project.

The author hereby disclaims any liability for injury to persons or property that may result while using this project book, and the author does not intend by this publication to explain all dangers known or unknown that may exist in the building and operation of this project.

Lindsay Publications has not built this project nor does it endorse the methods described. Lindsay Publications assumes no liability for injury to person or property that may result from the use of this information.

Write for a catalog of other unusual books available from:
Lindsay Publications
P. O. Box 12
Bradley, IL 60915-0012

LI'L BERTHA
A Compact Electric Resistance
Shop Furnace

text, photographs and drawings
by David J. Gingery

LINDSAY PUBLICATIONS

Li'l Bertha

A COMPACT ELECTRIC RESISTANCE SHOP FURNACE

by David J. Gingery

Copyright 1984 by Lindsay Publications Inc
Bradley IL 60915

All rights reserved.

ISBN 0-917914-16-3

1984

        11   12   13   14   15   16   17

For a fascinating selection of the highest quality
books for experimenters, inventors, tinkerers, mad
scientists, and a very few normal people... visit
www.lindsaybks.com

FOREWORD

Li'l Bertha is cute, compact, and she's a hot little number. Melting, forging, heat treating, burnout and calcining of investment molds, carburizing, tempering, enameling, firing ceramics, roasting and many other jobs can be done with ease.

"Li'l Bertha" was developed for use as a crucible furnace for melting aluminum, zinc die-cast alloy, and other low temperature alloys. It was developed in response to needs expressed by small shop operators at high schools and in plant maintenance operations where an occasional casting is needed or where moderately high temperatures are needed for a variety of purposes.

The name, "Li'l Bertha", was chosen from a list provided by my good friend, Nick Tyler, of Manotick, Ontario, Canada. "Li'l Bertha" is cute, compact, and she's a hot little number. Whether these attributes are shared by any ladies in Nick's present or past has neither been confirmed or denied, and I don't think we should ask.

Bertha is a resistance element furnace. Figure 1 shows a coiled resistance element of nickel chrome alloy. It is exactly like the filament of a light bulb except for its size and composition, and it radiates heat to the furnace chamber as it resists the flow of electrical current. This element has made it remarkably easy to build a high performance furnace—one that offers itself admirably for so many jobs that it is really quite difficult to give it a single name. Melting, forging, heat treating, burnout and calcining of investment molds, carburizing, tempering, enameling, fir-

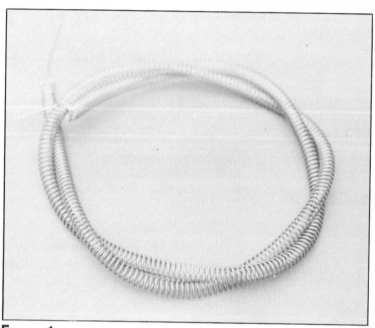

Figure 1

ing ceramics, roasting and many other jobs can be done with ease. It is so economical to build that the cost might be considered trifling, so there is really nothing to prevent you from having as many as you can use for a variety of purposes.

Since it burns no fuel, there are no products of combustion, and this is its main advantage over fuel burning units. Ventilation is a problem only if the substance being heated gives off gas, vapor or an odor that is dangerous or obnoxious, so the unit can be installed and operated anywhere adequate electrical power is available. A resistance element drawing as much as 1800 watts of power is still within the normal capacity of most ordinary household appliance outlets, so it is difficult to conceive of a shop where the unit could not be used.

Some basic design and electrical data are furnished, but it is entirely possible to proceed even if you have little understanding of the more technical aspects of electricity. Use the data provided mainly as an aid if you decide to modify the basic design for a specific purpose.

It is important to understand THERE IS some danger. Don't hesitate to seek help if you have any doubts about the safety of your installation. The obvious danger of electrical shock is added to the hazards of handling molten metal and other "hot stuff", so be absolutely sure that you are fully protected before you "plug in".

FIGURE 2

The basic unit uses what is called an "infinite range" surface unit control. One is seen in figure 2, and you may recognize it as identical to those on the family electric cooking range. It is easily obtained as a replacement part for domestic ranges, and it is also sold by dealers in repair parts for ceramic kilns as a "stepless heat control". It does a smooth job of controlling the current to the element from 5% to 100% of full capacity with practically no waste of power outside the heating chamber.

Information is offered to enable you to explore possibilities far beyond this simple design. More precise control and higher temperatures are quite easily had, so this project may well lay the foundation for a greatly expanded shop experience. It is certain that you can enjoy very substantial cash savings by building such equipment yourself.

THE DESIGN CONCEPT

1

Variations on the basic design are almost limitless. Your own needs may suggest forms that would occur to noone else. Simply consider basic limitations of the materials and proceed with reasonable caution. Cost is so slight that little will be lost if your idea doesn't work out.

The essential consideration in any project that is designed for home construction is that it must be constructed of easily obtained materials and within the technical capability of the average "do-it-yourselfer". The resistance element furnace is ideally suited for such a project because it is not even as complex as the family toaster. Construction materials can be obtained nearly anywhere at moderate cost, and much of it can be found as salvage at little or not cost at all. It is certain that no great amount of manual skill is required either, so persons of ordinary ability should not feel in the least intimidated.

THE FORMS

You've seen the heating element and one simple control in figures 1 and 2. With these on hand you'll need only a convenient mounting and housing for them to have a complete electric furnace. Simple sheet metal forms are shown in figure 3. These are made of light gauge galvanized steel, such as 28 gauge roof flashing material, and they are tamped full of a castable refractory concrete. When cured, the heating element and control is installed, and the furnace is fired with its own heating element to vitrify the lining.

1

FIGURE 3

These forms provide the base, body and lid of a simple crucible furnace. You probably already have guessed that the rubber tubing that is wrapped around the internal form will later be removed to provide the spiral groove for the resistance heating element. No special tools or equipment of any kind are needed, for the forms are merely sheet metal bands fastened together with screws. A hem or wire edge is easily put on the top and bottom of the forms with ordinary hand tools. Few shops today are without an electric drill, but even the hole drilling can be done with a hand drill if need be. Some scraps of pipe or tubing, a little plywood and some cardboard will will just about make up the rest of the required materials. A castable concrete product will be the best material with which to fill the forms, but even this can be made of home brewed materials for some classes of work.

VERSATILITY

Simplicity of design is probably this furnace's greatest asset. You can change the shape and size of

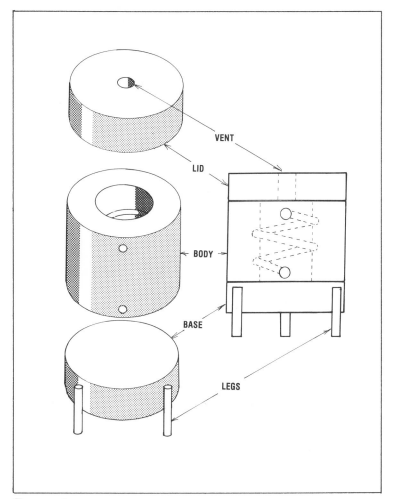

VENT

LID

BODY

BASE

LEGS

FIGURE 4

any of the parts to give the unit a different purpose. The basic shapes of the base, body and lid is seen in figure 4, and even these can be rearranged in various combinations and positions to meet special needs.

Figure 5 shows a profile of a unit with an added heat chamber to double the size of the unit. The extra chamber may or may not have a heating element, and it might be either taller or shorter than the basic body unit. If it does have a heating element it may be

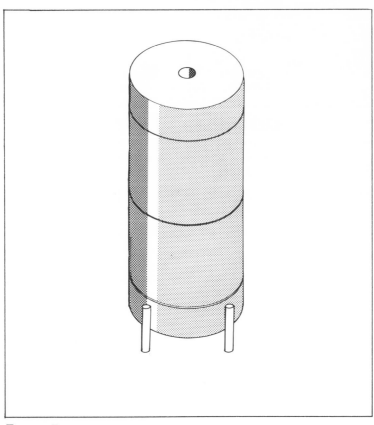

FIGURE 5

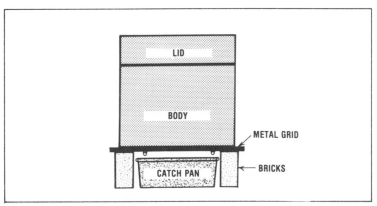

FIGURE 6

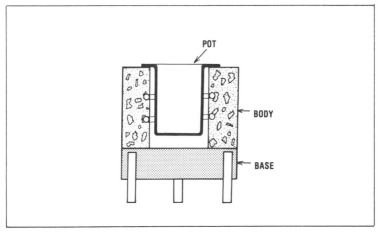

POT

BODY

BASE

FIGURE 7

operated without a variable control since the basic element can be moderated. The range of shapes that might be given to the chambers is limited only by your imagination and ability to construct them.

If your field of interest is lost-wax investment casting, you may want to provide a base with a grid so that wax can be melted from the molds. A pair of bricks upon which to rest the furnace body will do the trick. A metal grid to support the mold and a pan to catch the wax drippings will complete the set up. The final calcining of the mold can be done with the furnace's normal base unit.

Another popular need is for a simple dip furnace or pot furnace to provide a supply of molten zinc alloy, Babbit, type metal or lead. Just weld up a dip pot of standard steel pipe and weld on a flange to rest on the furnace body. Of course it is of special importance to insure that the pot does not contact the heating element, and that the pot and any part of the furnace that is exposed must be grounded. One variation of a dip unit is seen in figure 7.

Variations on the basic design are almost limitless. Your own needs may suggest forms that would occur to no one else. Simply consider basic limitations of the materials and proceed with reasonable caution. Cost is so slight that little will be lost if your idea doesn't work out.

HIGHER TEMPERATURES ARE POSSIBLE

The basic design assumes a high temperature of 1800° F. because that would fill the needs of most small shop operations. You can actually use a nickel chrome element at temperatures up to 2100° F, but the life of the element is noticeably shortened with prolonged use above 1800° F. The cost of replacing the element is so slight as to be considered negligible, but it may fail at a critical moment in an operation. If you must operate at higher tempereratures for prolonged periods you should obtain a heavy duty element from a dealer in repair parts for ceramic kilns. They are noticeably more costly, but they are rated up to 2300° F, which means that you can reasonably expect to melt brass in your furnace.

CONTROL OPTIONS

There is actually no need for any control of the element current other than a double pole switch if you plan to use your furnace only for melting, but some means of limiting the heat during the initial firing of the lining is quite necessary. Other operations require control too, and it is so simple to add the infinite range control that it makes little sense to be without the added advantage. An exception would be an extra body section to be used with the base unit. Either a series resistance control or the infinite range control are the only two practical options for most of us in the home shop. The application of automatic control by thermostat is not practical above 650° F, because such controls are too complex and costly. Even a simple pyrometer is beyond the reach of most of us, so we must devise ways to measure

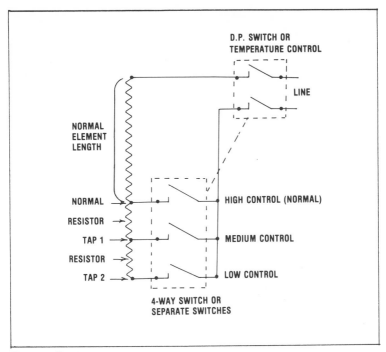

FIGURE 8

and control temperature without such exotic conveniences. This is no serious problem, for anyone can quickly learn how to do it.

SERIES RESISTANCE CONTROL

A rheostat is a variable resistor that is normally installed in series with an electrical load to reduce the voltage, and thus the current, to the load. It will usually be the first control idea that presents itself for such an application as the electric furnace, but it will prove impractical because such a device is simply too costly even if you are able to find one of adequate capacity for the work at hand. Also, a rheostat or any other form of external resistance will waste power as it radiates heat outside the furnace heating chamber. Obviously, it is better to apply the series resistance inside the heating chamber where any heat that is generated will be added to the heat of the main element.

An additional amount of coiled element can be added to the standard element to increase its resistance. One or more taps can be brought out to be a control panel to provide a variety of heat ranges in the same unit. Each tap can be connected to the line through separate switches, or you can get a heavy duty rotary three-way or four-way switch from a dealer in repair parts for ceramic kilns.

A schematic diagram of a unit with two heat ranges that are lower than normal is shown in figure 8.

Keep in mind that such a unit will operate at a lower peak temperature than one with a normal length element.

HIGHER THAN NORMAL HEAT

A standard element can be shortened to reduce its resistance value so that the increased current will provide higher temperature. Standard elements are usually designed so that they will operate at a non-destructive temperature where all of the heat produced can be radiated to still air. Most appliances use resistance elements in this way. As much as 10% of the original length of most elements can be safely removed, making the element run hotter. Although it will still operate below a destructive temperature, its life may be shortened if used this way for long periods of time.

I have not done so, but I have heard reports of brass being melted with nickel chrome elements. I suggest that you obtain a Kanthal or other heavy duty resistance element if you plan to operate above 1800° F for any prolonged periods of time.

A schematic diagram of a unit using both higher than normal and lower than normal heats is shown in

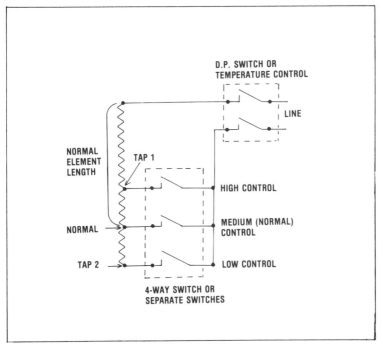

FIGURE 9

figure 9. Of course the same control options are open as for the example in figure 8.

Keep in mind that the peak temperature of this furnace will be both higher and lower than one with a normal length resistance element, and it is always possible that current could exceed the rating of controls.

JUDGING TEMPERATURE WITHOUT INSTRUMENTS

Most of us must operate our shops without the aid of accurate instruments because we can't justify their cost. But to measure high temperatures we don't really need an expensive pyrometer. It is fairly easy to estimate the temperature of heated objects by observing the color of light that is radiated. Possible errors due to difference of perception and amount of light in the surrounding area must be considered. You should assume that your estimate may be wrong by as much as 200° F, but that is usually close enough for most home shop operations.

Assuming there is little light other than that emitted by the glowing charge in the furnace, you can judge a dull red glow to be from about 950° F to 1000° F. Thereafter, as the temperature climbs, the red glow will brighten noticeably at about each 100 degree increment until it changes to orange at about 1600° F. The orange glow brightens through about 1900° F where it begins to show a yellow tone. It will be quite yellow at 2100° F, and it will show white at about 2400° F. It will be dazzling white at about 2600° F.

Standard charts have been printed that show the range of color change through various steps, but your own observation is likely to be more valuable than any charts. Those I have seen differ greatly in printing quality, and declared temperature changes vary from text to text.

Although this is a crude measuring system, you can quickly learn to judge temperature with reasonable accuracy for basic shop purposes.

The non-radiant, or tempering colors are equally useful, and you will find frequent reference to them in toolmaking texts. Again, consistent and accurate judgement will require practice. I see little point in telling you to heat a piece to a light straw color and then quench it, when we probably can't agree on what straw color is. A few experiments with a scrap of steel will teach you more than an entire book of conversation on the topic. Just observe as you heat and cool your work, and compare your observations with values that are known to you. Molten lead, tin and combinations are good standards with which to compare. The melting points of many combinations are widely published, and can be of great use in refining your judgement skill.

Application of the radiant color system will allow you to determine at what point you can tap a standard element to obtain the highest possible heat without destroying the element. Simply connect the rated voltage at both ends of the element at its standard length and observe the color. Then attach one of the leads to a point a short distance closer to the lead still attached to the end of the element. When you apply electrical power, the element will quickly heat through orange and change to yellow. This tells you that you have neared the limit of endurance, which is about 2100° F for nickel chrome. If the element does not heat to yellow, then you should again move the lead slightly closer to the end lead. You have to move the lead back and forth a number of times to find exactly the right point where the element glows yellow. You can then install the element with a tap at that point so that you can use it above its rating for brief periods when higher temperatures are needed. This test would be appropriate in the event that you use the method shown in the schematic plan of figure 9.

INFINITE RANGE CONTROL

Used is the standard range top burner control seen in figure 2, and I regard it as the most practical method of current control for this type of equipment. It is also used in some commercially-built ceramic kilns, where it is termed a "stepless control".

This device is not a thermostat, even though it may seem to perform as such on your kitchen range. It controls the current by varying the on-and-off time of the range element. Because the cycles are brief, color does not change perceptibly. The total length of each cycle may be from 20 to 30 seconds, and the on time will vary from about 5% at the low setting to 100% at the high setting. A mid-range setting may be about 15 seconds on and 15 seconds off, which will be clearly perceptible on the light-weight coiled element in your furnace.

The control has a small resistor that is wrapped around a bimetallic arm which controls a double-pole switch. The dial opposes the tension of the bimetallic arm to establish the setting, and current flows through the resistor in the control when the element is on. As the bimetallic arm heats and bends, it overcomes the tension of the dial until the current is stopped to both the element and the resistor. When the arm cools, it straightens out, closes the circuit, and the heating begins again. The overall effect is that the flow of current is smoothly moderated to reduce the amount of heat produced. No power is wasted as would be the case with an external resistance element.

This type of control can be added to any of the schematic diagrams seen in this book so long as the total current that passes through the control does not exceed the control's rating. Its capacity will be marked on the control, and all that I have seen are rated at 15 amps. This means that you can't exceed 1800 watts with a 120 volt unit, or 3600 watts with a 240 volt unit. Neither can you use the controls on voltages that differ from their marked ratings. The controls are extremely sensitive to overload, and will be destroyed in an instant by a short circuit.

It is likely that you may find these controls on a junked kitchen range, and the markings on the terminals can be identified from the diagrams in figure 10. These are the most commonly found controls, although some older types might be much different. I can't help you with them, but the appliance diagram should identify the terminals for you. The voltage and current ratings are especially important.

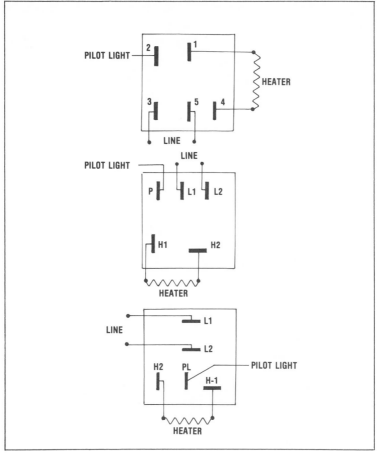

FIGURE 10

SAFETY CONSIDERATIONS

The additional danger of electrical shock and fire from overload along with the normal danger of operating a high temperature furnace calls for special caution. It is easy and inexpensive to provide protection. Be certain that you have done it right BEFORE you turn on the power.

You will find voltage and current ratings on the label or package of every electrical device that meets universal safety standards. Do not exceed the ratings

13

or do anything to frustrate the obvious intent of the basic design. This means that a three-pronged plug needs a three-pronged socket so that all terminals are properly connected to the line. Cutting away one prong or bending the prongs to make the plug fit into a socket not designed for it will destroy the safety it provides. Whether for 120 volt or 240 volt sevice, a third prong is provided as a safety ground. It ensures that current will pass to ground in the event of an accidental short circuit, and it is there to protect you from electrical shock. The metal shell of the furnace can become electrically charged at any moment without your knowledge, so make sure that the line cord is properly grounded and that all exposed parts are connected to the line ground.

Determine the maximum amount of current that your furnace will use, and make sure that the wiring in your shop can safety provide it. It is the size of the wire in the circuit that determines its capacity, and not the fuse or circuit breaker. Don't install heavier fuses or circuit breakers if your are not absolutely sure the wire is of adequate size to handle the larger current load. It is almost certain that the circuit will have been fused at its maximum rating at the time it was installed. Fusing a circuit at a higher rating than that for which it was designed is very dangerous. Don't do it.

Number 14 wire will normally be protected with a 15 amp fuse or breaker, but it is not safe for 20 amp service. Number 12 wire will be protected with a 20 amp fuse or breaker, and a 30 amp circuit will normally be of number 8 or 10 wire. The distance from the service entrance to the branch outlet enters into wire size calculations too, so give plenty of attention to the matter before you make any installations or changes in the present service.

The basic unit with a 1200 watt element will draw about 10 amps on a 120 volt circuit, so we are not speaking of unusual loads here. These are cautions

14

that should be observed in all of your daily use of electrical service in your home and shop.

Think through each operation before you begin. How will you handle the heated charge when it reaches its intended temperature? If it is a molten charge, is there something in your path to trip as you carry it? What if you drop the pot? Is there any combustible or explosive vapor in the area? What preparations and precautions are required to ensure safety and success of the operation?

The best idea that I know of is to prepare a written checklist and follow it as though you were working for a safety fanatic who will insist upon your full compliance to his rules of safety. Remember, we do these things for fun. If you get careless, you can burn down your shop and home, burn yourself, and possibly lose your life! This is supposed to be fun. So be careful!

THE REFRACTORY LINING

A refractory is a substance that can withstand high temperatures without being destroyed. The furnace heating chamber is lined with refractory which acts as an insulation so that as much heat as possible can be concentrated on the charge. There is a wide and confusing variety of commercially pre-mixed refractory materials available. But it is possible to blend your own refractory materials, so it would be worthwhile to study the topic.

For the purposes of this project we'll discuss only commercially blended products, although you may be able to produce your own unique castable refractory concrete from materials that are available in your area.

After testing hundreds of home brewed blends and some commercially prepared products, I finally selected Kaiser's "IRC" insulating castable refractory. It is rated at 2500° F, and it conducts heat at about 40% of the rate of ordinary firebrick. It is

15

FIGURE 11

noticeably lighter than the non-insulating castables. About 85 pounds will fill one cubic foot. It is available in 50-pound bags, which is adequate for small to moderate-sized furnaces. It sets up firmly in a short time and will endure rapid firing on the first heat. It is extremely strong in spite of its light weight. Other refractory manufacturers make similar products, though, to don't feel obligated to use this one product. Select the best refractory from those available to you, and you may find one even better. I have tested only a half dozen commercial products and there are a great many yet to try.

COMBINE ALL DATA

The shape, size and temperature range of your unit will be determined by your individual needs. It is highly likely you will be able to design a custom furnace that delivers economy and high performance yet is easy and low cost to build. The cost of producing a single unit is so low that you won't feel badly if it does not meet all of your needs on the first try. Two or more units would be a welcome addition to most of the shops I've seen. The second furnace should be easier to build, and should be of better quality because of the lessons you've learned in building the first.

The three units seen in figure 11 are sometimes found operating simultaneously in my shop. Their characteristics and capacities differ even though they look very much alike.

BASIC CALCULATIONS

2

Purposes other than melting aluminum or pot metal may present very specific heat requirements, so the basic formulas presented in this chapter can be of considerable value...It is assumed that you have absolutely no instrumentation to measure voltage, current or resistance, and these formulas are presented to aid you in judging the values without the use of instruments.

Unless you have a very specific purpose in mind it is likely that you can select a size from the table of basic designs and proceed to build with no pencil work at all. You may want to modify the design though, and it is possible that you might use these principles to design a much better furnace. Only the most basic design principles are presented here, but they should be adequate for average small shop needs.

UNIT SIZE

The primary use of the furnace is as a melting furnace, so it would be wise to build it to accept standard crucible sizes. Listings are available from crucible manufacturers and their dealers, but smaller sizes will be listed here as a partial guide.

Size numbers of crucibles indicate their approximate capacity in molten aluminum, and they will hold about three times as much weight in zinc, die cast alloy, or brass. A number 1 crucible will hold about one pound of aluminum or about three pounds of brass or pot metal, and a number 2 twice as much, etc.

Size Number	Height	Bilge Diameter
1	3 5/8''	3 1/8''
2	4 1/2''	3 11/16''
4	5 3/4''	4 9/16''
6	6 1/2''	5 1/4''
8	7 1/8''	5 7/8''
10	8 1/16''	6 9/16''

Larger sizes are available, but are not generally used in the home shop. A number 4 will be adequate for most home shop operations.

Add about 1 1/2'' to the bilge diameter of the crucible and about 1/4'' to its height to determine the heating chamber size for the melting furnace. The cubic content of a cylinder is found by multiplying the square of the diameter by .7854, and then multiplying that product by the height of the chamber. If diameter and height are measured in inches, then the volume calculated will be in cubic inches.

Unit Size	Height	Diameter	Cubic Content	Watts
1	4 1/4''	4 1/2''	65 cu in	1200
2	4 3/4''	5 1/4''	103 cu in	1800
4	6''	6 1/4''	184 cu in	3600
6	6 3/4''	6 3/4''	242 cu in	3600*
8	7 1/2''	7 1/2''	331 cu in	3600*
10	8 1/2''	8 1/2''	482 cu in	3600*

* Larger elements can be used in these large units, but the infinite range control will handle only 15 amps. A dual element with compound controls will be required for more rapid heating in large units.

The cubic content of these units is given for comparison in the event that you use a chamber of different shape. You might reasonably assume that if a furnace of 65 cubic inch content will handle 1200 watts, then one of 130 cubic inches would handle 2400 watts. This is basically true, but other factors enter as well.

A thicker wall will retain heat longer, and thus an element may overheat. Experiments would have to be performed to determine the right combination of chamber size, refractory thickness, and wattage. I have done no such experiments to date.

The applicable chamber sizes allow for standard crucibles, clearance for crucible expansion, and clearance for a set of tongs to take the crucible out at the end of the heat. It is best to have both the pot and the tongs on hand before you start design and construction.

CRUCIBLE HANDLING

You can make a satisfactory melting pot by welding a disk of steel to one end of a length of standard weight steel pipe. A stroll through the local junkyard will provide the material, and a local welding shop will do the work for you at a moderate cost if you don't have a welder. I recently had five of them made for less than $20.00, and that is less than the cost of one clay graphite crucible. Such a steel pot can be easily handled with tongs.

There is some work though that will require a ceramic pot. Silicon carbide crucibles are very durable, but quite costly. Clay graphite pots are about half as expensive, but they are likely to break if not carefully tempered before their first use. The electric furnace is especially well suited to ceramic crucible melting because you can control the heat so much better than with fuel-burning units.

Simply raise the temperature slowly with the first heating until the pot is a bright red. Turn off the power, and let it cool slowly in the furnace until the next day. Then you can put in the charge and proceed to melt.

You can order crucibles and other foundry supplies from Metal-Max, PO Box 3673, Oak Park, IL 60303.

The tongs that are used to grip the pot must be made to fit very accurately. They must grip just below the bilge, and must never apply pressure to the rim. If gripped too tightly or improperly, a crucible can burst like an egg and drop a molten charge on your feet. A metal pot can be gripped by the rim, but a ceramic pot should not be handled in this way except for very small laboratory crucibles. This is very dangerous work, and you must address it very seriously.

HEATING ELEMENT SIZE

Applications other than the melting of aluminum or pot metal may present very specific heat requirements, so the basic formulas presented in this chapter can be of considerable value. The rate of temperature rise in the furnace is a direct result of heat being generated faster than it is absorbed by the load or faster than it escapes through leaks and by conduction through the lining. It is possible that an 1800 watt element could be used in the number 1 furnace, but it may overheat and burn out if the furnace load is not large enough. Conversely, an 800 watt element may eventually melt a charge of aluminum in the same furnace, but it would take much too long. Experiments and calculations should make it possible for us to select an element size that gives the highest possible heat, but that can be moderated to do the work in any range below the maximum.

HEAT CALCULATIONS

Heat is a form of energy, and it can be measured and discussed positively and precisely even though it, in itself, is not visible or tangible because we can observe the effect of heat on visible and tangible objects. It helps to clarify matters when we can agree that "cold" is really the absence of heat, for our goal is to generate, contain, and concentrate heat so that it will do the work we need done. When it escapes, it is lost to the surrounding air, and we have paid for wasted electrical energy.

When heat is added to a substance, its temperature rises, and this is expressed in degrees, either Fahrenheit or centigrade. The difference in these thermometer scales is seen in that water boils at 100 degrees on the centigrade thermometer and at 212 degrees on the Fahrenheit scale. Having pointed out that there is a difference, I will say no more about it except to point out that I use the Fahrenheit scale for the same reasons that I measure in inches instead of centimeters. If your work requires that you use the centigrade scale you can surely do it without my help or hindrance, as the case may be.

There is a significant difference in the amount of heat that is required to raise the temperature of various substances, so another means of expressing heat value is the "BTU", which is short for British Thermal Unit. This is defined as the amount of heat energy required to raise the temperature of one pound of water one degree Fahrenheit. The scientific expression is actually much more precise than that because the amount of heat varies as the temperature of the water varies. That is, more heat is required to raise the temperature of water from 100 degrees to 101 degrees than is needed to raise it from 35 degrees to 36 degrees, but the difference is so slight that it need not enter into most calculations.

Specific heat is the ratio of BTUs needed to raise the temperature of a substance one degree to the BTUs needed to raise an equal weight of water on degree. If one BTU is needed to raise one pound one degree, the specific heat is 1. Specific heat tells us how much heat energy a substance holds at a given temperature. Because water has a much higher specific heat than lead, a pound of steam can produce far worse burns to your hands than can a pound of lead. Greater or lesser values of specific heat are expressed in decimals or combinations, and accurate tables are available in technical manuals. These values enter into the calculation of expected heat loads for specific materials.

The unit of electrical power is a "watt". Its value is calculated by multiplying the voltage in a circuit by current in "amps", and it is normally expressed in "kilowatts", which is 1000 watts. For example: a current of 10 amps in a 120 volt circuit would produce 1200 watts, or 1.2 kilowatts. (10 × 120 = 1200 or 1.2 thousand)

The quantity of heat that is produced by one watt of electrical power is 3.412 BTUs. Thus, one kilowatt will produce 3412 BTUs of heat. It is necessary only to multiply the electrical power rating of your element in watts by the factor 3.412 and you know the heat producing capacity of the furnace in BTUs. An 1800 watt element would produce 6141.6 BTUs. (1800 × 3.412 = 6141.6)

Formulas and tables have been developed so that you can accurately calculate the precise amount of heat that is required to raise the temperature of a given substance to its melting point and to superheat it to a practical pouring temperature. Such calculations are seldom used in the home shop, though, because we lack the equipment to accurately measure and control temperature. But, the formulas can be an aid in estimating the basic amount of work that might be done with the equipment that we have. The important thing to remember is that no furnace is 100% efficient, so if you calculate that your furnace will do a specific job in a specific time by these formulas you are likely to miss it by a mile or so. A unit with an efficiency rating of 50% would be pretty good, so double your estimate for an approximate figure and expect to do a bit better or worse.

ELECTRICAL CALCULATIONS

Electrical current, like heat, is not visible, but its

effects on tangible things are. Very precise physical laws govern its action in a circuit. Our application of these laws in this instance is quite simple because we are concerned mainly with the heat that is produced by resistance to the flow of current. Even though we use standard alternating current, our calculations use direct current formulas because AC voltages and currents are measured in such a way that their effect on circuits is the same as the equivalent DC voltages and currents (RMS measurements). It is assumed that you have absolutely no instrumentation to measure voltage, current or resistance, and these formulas are presented to aid you in judging the values without the use of instruments.

It is the resistance to the flow of current that causes electrical energy to be converted to heat energy in the coiled resistance element. The physical law that deals with electrical resistance is called "Ohm's Law", in honor of the man who first understood and expressed it. The law states that "The strength of a continuous electrical circuit is directly proportional to the voltage, and inversely proportional to the resistance." That is, the greater the voltage, the greater the current, and the greater the resistance, the lesser the current. Each factor has an effect on the others, and all are considered in each calculation.

The voltage in a circuit is the driving force, and it can be compared to the pressure in a hydraulic circuit, such as a water line in your home. Its symbol in electrical formula is "E", which is short for "EMF" (electro-motive force). The unit of measurement is the "volt".

The current in a circuit can be compared with the volume of flow in a hydraulic circuit, such as how quickly can you fill a 10 quart pail at the kitchen faucet? The unit of electrical power is the "watt", as previously explained, and it is the product of voltage and current in the circuit. The symbol for current in electrical formula is "I", and the unit of measurement is the "ampere" or "amp".

The resistance in an electrical circuit can be compared to a restriction in a hydraulic circuit, such as a pinched pipe or a partially closed valve. Its symbol for calculations is "R", and the unit of measurement is "ohm".

By using these values in a simple formula, we can calculate the circuit values and put intangible electricity to work. A great many problems can be easily solved by the application of this simple formula, otherwise known as "Ohm's Law":

$$I = E/R \qquad \text{- or -}$$
$$I \text{ (current)} = E \text{ (voltage)}/R \text{ (resistance)}$$

This is to say that current in the circuit is equal to the voltage, in volts, divided by the resistance in ohms, with the current being expressed in amps.

The standard algebraic law of transposition of terms applies, so: $R = E/I$, and $E = I \times R$. Using these simple formulas, it is easy to find any unknown value when the other two values are known.

APPLYING THE FORMULAS

The main reason for including these formulas and values is to aid you in building and understanding your electric furnace even if you don't have any type of electrical measuring equipment. Electrical meters can be quite costly, and it is not wise to borrow them because they can be easily destroyed by improper use. You can come quite close to actual circuit values by calculations without having to use valuable test instruments.

Keep in mind that engineering calculations are really little more than an educated guess. Factors not considered may be of greater importance than those considered in the calculations. It has been determined, for instance, by very precise calculations, that 517 BTUs will melt and superheat one pound of aluminum to a proper pouring temperature. Such a statement in a respected journal may lead you to conclude that since one watt will produce 3.412 BTUs, then you need only divide 517 by 3.412 to find that 151.5 watts will do the job. It sure does sound impressive. But wait! If that's true, then we can do the job with a couple of light bulbs and have power to spare. No need to build a furnace at all.

The calculations are based on proven data, so how can they be wrong? It is the unknown and unconsidered factors that are to blame. These calculations are based on such information as the specific heat of aluminum, the melting point, latent heat of fusion, superheat to pouring temperature, but not on the efficiency of the melting unit. When the drastic losses of the initial start up are considered it is likely that several times as much heat will be used before a pound of aluminum is melted. When the unit is hot, it may seem to do the job with half as much. With the proper respect for the validity of figures, we can proceed to determine some factors that are of interest and importance.

A 1200 watt element is listed in a catalog with a voltage rating of 120 volts, a coiled diameter of 9/32" with its unstretched length being 15". That's all you need to know if you plan to use it as is, but the rest of the information is there if you need it. Try these simple calculations just to see how easy it is:

1. Find the current: $W/E = I$ or
   1200 watts/120 volts = 10 amps

2. Find the resistance: $E/I = R$ or
   120 volts/10 amps = 12 ohms

It is important to know the amount of current so that the capacity of wires, controls and branch circuits won't be exceeded. Note also that it would not have been possible to determine the resistance without the current value. Knowing the resistance and current value will also be of help in determining other factors.

Knowing that the element draws less than 15 amps, you may want to shorten it to increase its operating temperature. By stretching it lightly between two pins on an asbestos shingle you can apply the rated voltage to various points along its length and observe the radiant color. Suppose you find that it will operate at an acceptable temperature when it has been shortened 1″. Now you will want to know the new current and watt rating of the shortened element.

3. Find the resistance:
   old length = 15″, new length = 14″
   therefore new resistance = 14/15 of total

4. 14/15 of 12 ohms = 11.2 ohms

5. Find the current:
   E/R = I or 120/11.2 = 10.71 amps

6. Find the new power:
   I × E = W or 10.71 × 120 = 1285 watts

Now you have determined the value of the element if it is used with a tap at 14″ of its length to give you a high heat of 1285 watts and a low heat of 1200 watts. You know that it is still within the capacity of a 15 amp cord (1285/120 = 10.7 amps), and that its control and other components need not be rated above 15 amps or 1285 watts.

Perhaps you would like to add another tap to provide a heat that is below the standard rating of the element, to make your furnace even more versatile. Let's say we'd like to add an amount of resistance element to give a low heat of 1000 watts:

7. Find the current: W/E = I or
   1000 watts/120 volts = 8.33 amps

8. Find the resistance: E/I = I or
   120 volts/8.33 amps = 14.41 ohms

The standard 1200 watt element is 15″ long, so each inch represents a resistance of .8 ohms. An additional 3″ of the element will give a total of 14.4 ohms because 18 × .8 = 14.4.

Now you have designed an element that will give heats of 1000, 1200, and 1285 watts at three taps, and you know that the total current draw of your furnace will not exceed the rating of its cord, controls, or branch circuit. You may also want to know the heating value of each tap in BTUs for such operations that require that knowledge:

9. Low tap - 1000 watts × 3.412 = 3412 BTU
   Med. tap - 1200 watts × 3.412 = 4094 BTU
   High tap - 1285 watts × 3.412 = 4384 BTU

(Note: The electrical resistance of electric heating elements often changes with temperature. A cold element might have a measured resistance of 5 ohms. at 1000° it might be 15 ohms, and at 1500° it could be 20 ohms. This is another unknown that could throw off your calculations.)

A schematic of the element design application is seen in figure 12. It can be controlled with an infinite

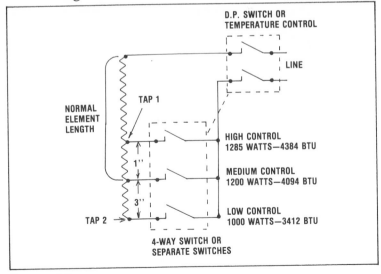

FIGURE 12

range control, individual switches, or a four-way switch rated at least 15 amps. The maximum temperature at each tap will differ noticeably.

It is possible that you may want to design with two or more elements that are separately controlled, so you will want to know the total current draw of all elements to ensure that the supply is adequate. For whatever purpose, you can use these simple formulas to find the realistic value of the element in its application, and you won't need expensive meters to do your work.

Standard resistance heating elements are generally furnished as a close-wound coil that can be stretched as much as three times its original length when it is finally installed.

Elements for most applications are designed to operate below a destructive temperature when all of their heat can be radiated to still air, but some special elements are rated for forced air applications.

Some factors of design and construction cannot realistically be calculated, so we are still left with an amount of guesswork. But our guesses are more educated now, and we have considerably more control over unknown factors through the infinite range control.

Standard elements are so inexpensive that it would really not make sense to invest a large amount of cash in a test meter merely to protect an element that costs just $3.00 or $4.00.

The basic model that is presented in detail is within the capacity limits of its control and a standard 20 amp circuit. It will operate at approximately 1600° F with no load in the heating chamber, and can operate for several hours at a session without oxidizing the nickel chrome heating element. It is specifically designed for melting aluminum in a standard clay graphite crucible or a steel or iron pot.

The efficiency of the furnace is greatly affected by the insulating value of the furnace base, walls and lid. A furnace would be 100% efficient if all of the heat generated were absorbed by the load, but of course, some portion of the heat will be lost through the refractory lining. In test sessions as long as 10 hours, a piece of dry paper in contact with the furnace body did not char or ignite. Neither did a piece of paper left under the furnace for the entire test period. Even so, the outside surfaces will get well above 200° F, which will cause a painful, if not serious, burn to the naked skin.

THE BASIC
MELTING FURNACE

3

Li'l Bertha can be used for many purposes and processes other than melting, but some processes are chemically active. No salt-glazed ceramics can be fired in a furnace of this type, and the possibility of a charge or load becoming chemically active upon heating should be known by you before the furnace is loaded.

FIGURE 13

A set of forms for a complete melting furnace are seen in figure 13. They are merely rings made of galvanized sheet metal, and the dimensions are easily calculated when you have decided on the final size of your unit. The bands of metal that form the rings are simply overlapped and fastened with screws or rivets; then the top and bottom edges of the base,

33

body, and lid are finished, either with a hem or a heavy wire rolled into the edge to provide a non-cutting edge. The unit seen in the photos is a number 1 size, but other sizes are made in the same manner.

The size of these forms is based on an inside diameter of 4 1/2'' and a height of 4 1/4''. The wall thickness is 2 1/4'', so the outside diameter of the unit will be 9''. Having these basic details on hand, it is a simple matter to lay out and form the dimple sheet metal rings that make up the forms and ultimately the outside skin of the furnace.

FIGURE 14

This is the first order of calculation and construction, and it is the easiest of all the forms. It includes a length of rubber tubing that is wrapped spirally

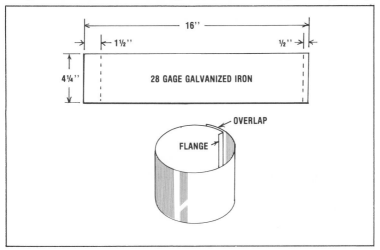

16"

1½"

½"

4¼"

28 GAGE GALVANIZED IRON

OVERLAP

FLANGE

FIGURE 15

around the form to provide a groove for the heating element in the main body. I found that a length of 3/8" OD vacuum hose from the auto supply store was just right for the form. It is just a tiny bit larger in diameter than the coiled heating element, and it leaves an ideally sized spiral groove when later removed from the body casting. Of course, a larger hose must be used if the element is to be larger. A view of the completed form is seen in figure 14. It is rubbed with paraffin to ensure easy release from the refractory material when it is removed.

The circumference or periphery of a circle is found by multiplying the diameter by 3.1416, so a form of 4 1/2" diameter will have to be 14.137" long plus an amount for overlap and a flange. Since the diameter is not critical, there is no reason why we can't round of the calculation to 14 1/8" or even a simple 14". A flange of about 1/2" width will stiffen the form where it overlaps, and an overlap of about 1 1/2'' will be adequate. Simply cut out a rectangle of light gauge galvanized sheet metal that is 16'' long and 4 1/4'' wide and bend up a 1/2'' flange on one end as seen in figure 15.

Cut a ring 4 1/2'' in dameter from 1/2'' plywood with its rim about 3/4'' wide. Cut a saw slot in the rim to accept the flange of the form.

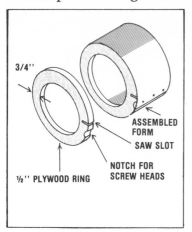

FIGURE 16

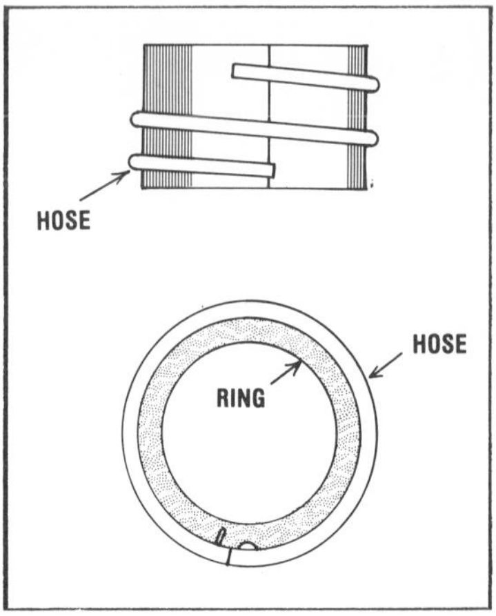

FIGURE 17

Drill three 1/8'' holes through the assembled form to receive small sheet metal screws. The screws enter the form from the inside, and they fasten the rubber tubing in place with only a slight tension. Install one screw through the form and the tubing, stretching the tubing slightly as you install the center screw. Stretch it a tiny bit more as you install the final screw. Your form will look that in figure 14. Note that there is no hem or wire edge on the inside form because it will not be a permanent part of the furnace.

The wooden ring is notched so that it can slide past the screwheads to give easy access to the screws. Phillips head screws are much easier to use in this application than slotted screws because a stubby screwdriver must be used from the inside of the form.

Since we have decided on a wall thickness of 2 1/4'' and an inside body diameter of 4 1/2'', the outside body diameter will be 9''. Multiply 9 by 3.1416 and we have a circumference of 28.274'', which can be rounded off to 28 1/4''. Add 3/8'' to each end for overlap and 3/8'' to both top and bottom of its 4 1/4'' height and you have a piece of galvanized sheet metal that measures 29'' x 5'' for the outside form as shown in figure 18. The band is rolled into a neat hoop and joined with three or more sheet metal screws or rivets.

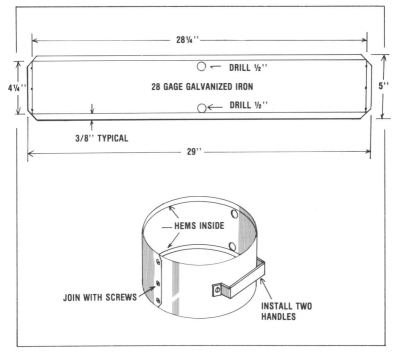

FIGURE 18

Make or purchase four sturdy handles. Two will be installed on the body and two on the lid. Such handles are available at hardware departments in a variety of styles. I chose a 6 1/2'' heavy duty pull handle and altered the contour of the mounting ears to conform to the curve of the body. A 1/4″ bolt on each end will be adequate to hold the handle.

Note that two holes of 1/2″ to 3/4″ diameter are made in the outside form to coincide with the ends of the tubing on the inside form. A pair of dowels will pass through these holes, and a small brad in the end of each one will enter the rubber tubing to hold them in position while the refractory is tamped in. The dowels are drawn out as soon as the refractory has set up firmly, and the holes left by them will provide entry for the terminals that supply power to the heating element. Of course, an element that is to be tapped at points other than at each end will need a hole for each tap terminal.

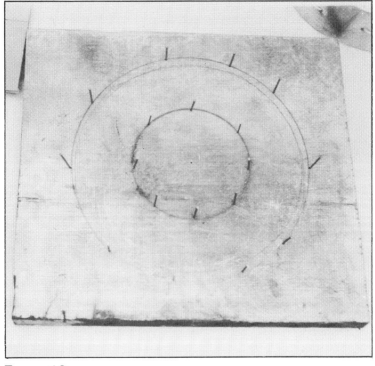

FIGURE 19

ASSEMBLING THE BODY FORM UNIT

A piece of plywood is marked with a pair of concentric circles that represent the inside and outside forms. Small brads are driven into the circles at a slight angle so that they will guide the forms to their proper positions as shown in figure 19.

Note in figure 20 that the wooden ring in the inside form is at the top to hold its shape, and the ring of brads in the plywood base hold the shape of the bottom of the form.

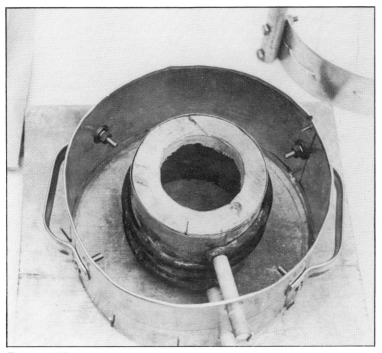

FIGURE 20

The outside form is held in position and true shape by its ring of brads, and the two dowels are slipped through the holes so that their pins enter the rubber tubing. Finally, a carriage bolt or length of threaded rod is installed through the center of the board and a bar or cleat is bolted snugly on top to hold the forms in place while the refractory is tamped in, as seen in figure 21.

The main body mold is now ready for its refractory lining, which is mixed with water and tamped in place according to special directions, which will be given a bit later.

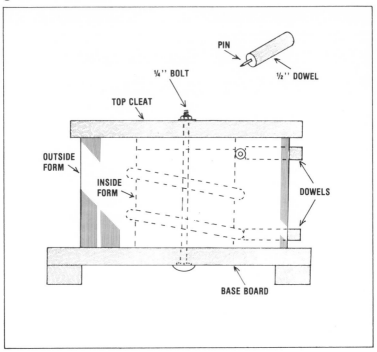

FIGURE 21

THE LID FORM

Most of the story is seen in figure 23. Its finished diameter, and thus the length of the form, is the same as the body form. Its height is the wall thickness of 2 1/4″ plus the top and bottom hem allowance of 3/8″ for a final dimension of 29″ x 3″, as shown in figure 22.

Note that the handles on the lid and the wire mesh that is woven through 16 holes are positioned

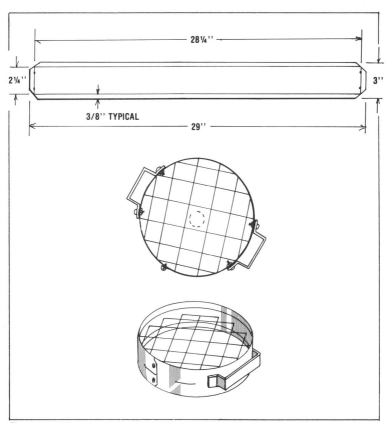

FIGURE 22

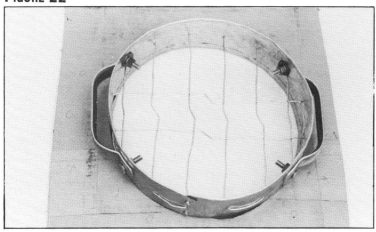

FIGURE 23

on the center line. The wire is about 24 gauge galvanized, and it is woven first loosely, then it's snugged by crimping it in the center when the form is set in a cardboard ring to hold its shape true round.

A wooden dowel of about 1 1/4'' diameter is set in the center of the lid form when the refractory lining is tamped in place to leave a vent. The finished lid form is shown in figure 23.

BASE FORM

The dimensions of the sheet metal form for the base are identical to that of the lid. The wire grid is installed in the same way too, but three legs are installed in place of the handles. The base form is shown inverted in figure 24.

Three legs are used instead of four so that the furnace will rest without wobble even if set on an uneven surface. Lengths of 1/2'' pipe or conduit about 5'' or 6'' long will make adequate legs. You might also use angle iron, as seen on the larger furnace, or square tubing, or any other shape that lies about handy. The object is to provide 2'' or 3'' of free air space beneath the furnace. The base form, like the lid, is set in a ring to hold it concentric when the refractory is tamped in.

The three parts of the furnace will look like those in figure 25 when the refractory lining is tamped in place.

THE REFRACTORY LINING

As mentioned earlier, there are a number of castable refractories available. The study of furnace refractories is in itself complex. An entire manual could be devoted to the topic alone, but it will have to be written by one more knowledgeable than I. Over 300 hours of experiments in my shop have convinced me that it is more practical to buy a prepared mix if the intended temperature range will exceed red heat

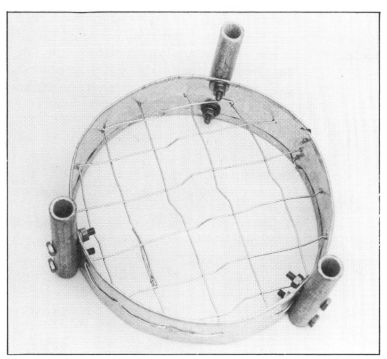

FIGURE 24

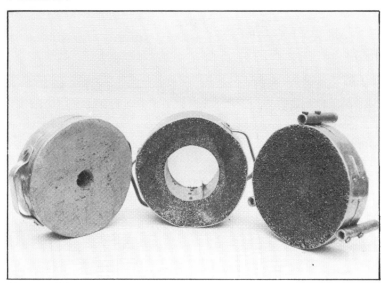

FIGURE 25

(about 1200° F). There are many conceivable operations where moderate temperatures are used though, so there is no reason why you can't blend your own refractory from common materials or use a commercial blend that is rated for a lower temperature. Here are a few thoughts to guide you as you consider your choice of refractory lining.

1. Most higher temperature commercial blends are stronger and more resistant to abrasion.

2. The higher temperature formulas are not usually as efficient insulators as the lower temperature formulas.

3. The blends that are described as "insulating" will usually be much lighter in weight than the non-insulating blends.

4. The vermiculite based blends show the best insulating quality, but they are very noticeably weaker in crushing and rupture strength, and not very durable in abrasion tests. These would appear to be good back-up insulation in a more complex furnace, but not the right material to support the heating element, or withstand the abrasion and shock of daily use in contact with the charge or load.

5. Not all varieties of any given manufacturer are likely to be available from every dealer. A compromise is likely to be necessary, so make a comparison of available materials, not what might be listed in a comprehensive catalog. It is unlikely that you could order just one or two bags of a non-stock item from a dealer.

6. After experimenting with other brands and a large number of home-brewed mixes, I chose Kaiser's "IRC" insulating castable refractory. It is rated at 2500° F, and its density is 80-85 pounds per cubic foot. Its thermal conductivity is rated as approximately 40% that of firebrick. I found it very easy to use. It cures to a very durable lining with little tendency to crack or shrink. The 50-pound package size makes it very convenient for small shop use. The same com-

pany offers a broad range of other refractory types and grades as well, so do have a look at what is available locally before you make a choice.

7. A castable refractory is in reality a heat-resistant form of concrete, the ingredients of which are likely to include fire clay, some grade of aggregate or grog, and a green strength binder. There are countless formulas. Some differ only in the proportion of the same ingredients, and vastly different materials result from seemingly slight changes. Most of the materials used in refractory mixes are available to the ceramic or pottery hobbiest. If you are not familiar with the handling and curing of ceramics it will likely be best that you choose a commercially blended product, and follow the directions that are provided with the product.

PREPARING THE REFRACTORY MIX

Not all packages have complete directions, so note these general principles:

1. Use a clean container and clean water to prepare the mix.

2. Old mortar, lime or cement are likely to weaken the mix or to cause rapid setting or curing problems.

3. Use only enough water to bring the mix to proper working consistency. Excess water weakens the product, causes excessive shrinking, and may seriously affect the setting and curing time.

4. Most castable refractory products begin to set in about 30 minutes. You should plan your work so that all material mixed will be in the forms within 20 minutes. Water added to make a semi-set mix workable will seriously weaken the product. No troweling should be done after the initial setting period has passed.

5. Forms should remain in place for 24 hours before removing them. To remove the forms before

ultimate strength is reached may destroy all of your work. Be patient. Let it set at least overnight.

6. The 24-hour set-up period is considered curing time. It is generally best to cover the work with plastic or a damp cloth during the initial curing time.

7. When the curing period has passed, the material is then dried by gradual heating before it is raised to a red heat to vitrify it. It's best to avoid any attempt to air-dry the cast material because that increases the danger of failure from rupture or spalling. Trapped water may form steam pockets, and a casting will explode before it is dried out.

8. The usual rate of heating is to increase the temperature to 250° or 300° F in 50° F steps per hour. The temperature is held there for 4 to 6 hours, and then raised to 500° or 600° F in 50° F steps per hour. As a final step the temperature and the power is turned off to allow the lining to cure as slowly as possible. Since some cracking and damage to the lining is bound to occur in daily use, I decided to speed up the initial firing to include all heating steps in a time space of about 6 hours. I then let it cool slowly overnight. The only damage seemed to be small radial cracks that didn't appear to impair the operation of the unit in any way.

9. The material for the lid and base was mixed until quite stiff because it did not have to conform to the fine detail of the rubber tubing. The material for the body was mixed with enough water so that it was easily vibrated into close contact with the tubing by tapping the forms. I used each form as a measure to prepare the proper amount, and added a small amount over the dry volume to be sure there would be enough when blended with water.

INSTALLING THE HEATING ELEMENT

When the screws are removed from the inside form, it will collapse and easily slip out of the body.

The rubber tubing will still be in place. When it is pulled out, as seen in figure 26, it will leave a spiral groove into which the heating element is installed. This is done after the form has set for 24 hours after casting, when the mix has reached its full strength.

FIGURE 26

The terminals that anchor the ends of the heating element and carry its connections to the outside of the body are best made of number 8 or number 10 stainless steel machine screws, but they can also be made of ordinary steel screws or threaded rod if stainless steel is not readily available. In either case, it is likely that the terminals will have to be replaced each time the element is replaced, because the nuts will usually be impossible to remove due to scale build-up. Two possible terminal designs are seen in figure 27.

The wooden dowels were withdrawn from the body as soon as the refractory mix attained its initial

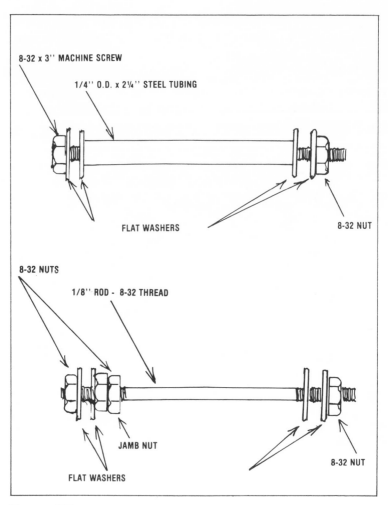

8-32 x 3'' MACHINE SCREW

1/4'' O.D. x 2¼'' STEEL TUBING

FLAT WASHERS

8-32 NUT

8-32 NUTS

1/8'' ROD - 8-32 THREAD

JAMB NUT

8-32 NUT

FLAT WASHERS

FIGURE 27

firm set. The holes meet the ends of the spiral groove left by the rubber tubing. You can prepare a small amount of fire clay to the consistency of stiff putty, or you can buy prepared ceramic clay from a dealer in ceramic pottery supplies. The clay is pressed into the terminal holes with a length of dowel until it is completely filled; then the terminal is pushed through the clay, centered in the hole and imbedded in the clay to

48

insulate it from the sheet metal skin of the main body. Although the wooden dowel is removed as soon as possible after the initial set, the terminal is not installed until the inside form has been removed and the body has set for 24 hours. An enlarged cutaway view of the terminal installation is seen in figure 28.

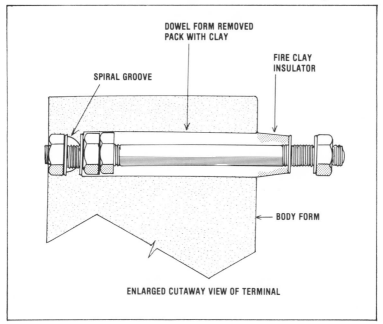

ENLARGED CUTAWAY VIEW OF TERMINAL

FIGURE 28

Note that a mound of clay about 3/8" high is formed outside the body so that the terminal connections to the control and the line are insulated from the body skin. The element and control are installed while the clay insulator is yet pliable and damp. It hardens and dries during the first firing of the furnace.

A small loop is formed in each end of the heating element, and is stretched uniformly until it is slightly longer than the rubber tubing that formed the groove. One end of the element is connected to one of the terminals, and then it is slipped into the groove

FIGURE 29

until the other end can be connected to the other terminal. The element should be just a bit too long so that it is forced slightly against the bottom of the spiral groove when the other end is connected. After the first heating, the element will have expanded and contracted so that it will always be a bit loose in the groove when it is cool, but it will fit nicely when red hot. The element is seen installed in figure 29.

INSTALLING THE CONTROL

The flexible supply cord must be firmly clamped so that it won't put pressure or tension on the terminals of the control or furnace, and so that it can't be accidentally pulled loose during operation. Brackets of nearly any size or shape can be made if you have equipment to work in heavy sheet metal. Standard electrical box covers with a knockout for a clamp

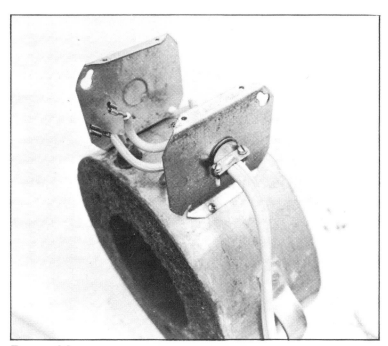

Figure 30

Figure 31

51

FIGURE 32

make handy brackets to support the cord and the control cover. The ones I used are intended for a standard 3 1/2'' octagon box, and I simply bent a 3/8'' flange on two opposite sides and drilled holes for mounting them to the furnace body with sheet metal screws. The two prepared brackets with the 14 gauge flexible cord are seen in figure 30.

Note in figure 31 that the flexible cord has its third wire connected to the sheet metal skin of the body with a sheet metal screw and washer, and the other two ends have 1/4'' solderless terminals to fit the spade terminals on the control. The two wires that connect the control to the terminals of the furnace are of number 14 asbestos-covered range wire which can be stripped from a scrapped cooking range. Tight connections can't be made to the furnace terminals while the clay is yet soft, but they can be snug enough for initial firing of the furnace lining.

FIGURE 33

The complete wiring of the control and element are seen in figure 31, and the cover is seen installed on the brackets in figure 32.

THE INITIAL FIRING OF THE LINING

As mentioned earlier, it is best to do the initial firing soon after the curing period, to avoid spalling due to trapped moisture in the lining. The first phase of the initial firing is to drive out water in the form of steam. When you first turn it on you will hear hissing at the terminals and steam is likely to be very actively escaping from the soft clay insulators. This will go on for some minutes, but it will soon quiet down to a gentle heating and a gradual escape of vapor as the temperature of the refractory lining increases. The cured base is set on a non-combustible surface, and the body is set in its normal place on the base. The lid is set in place as soon as proper heating of the element is proven, but the vent hole is left open until all of the steam has been driven out of the lining. Then the vent hole can be covered while the furnace runs to full heat for an hour or so.

The first two hours of firing should *not* be above the number 2 setting on the infinite range control dial. The next two hours can be at the number 4 setting. By the end of the first four hours it is likely that the lining will appear completely dry, but all moisture will not have evaporated until it has reached a red heat. The third two-hour period of firing can be at the number 6 position on the dial, after which you can cover the vent and move the control to its highest setting for an hour or two. By the end of the eighth hour, the inside of the lining should be about 1700° F with a standard element at full operation, and the outside skin will be in excess of 250° F. The lining will probably be radiating bright orange light to a depth of 1/2'' or more, and it is unlikely that any water will remain in the lining material. It's best to leave the lid in place with the vent hole covered, to allow the furnace to cool slowly overnight.

Small radial cracks appeared in my main body lining after the first firing, but they have not increased in size or number since that time. I have exceeded 100 hours of testing and some of the sessions have been as long as 10 hours with no apparent damage to the element or lining.

LIMITATIONS

While I have not done so, I've heard reports that brass can be melted with an element of nichrome V wire. It can certainly be melted with an element of Kanthal wire, which is rated at 2300° F.

The furnace can be used for many purposes and processes other than melting, but some processes are chemically active. No salt-glazed ceramics should be fired in a furnace of this type, and the possibility of the charge or load becoming chemically active upon heating should be known before you use the furnace for such a purpose.

As a safety precaution, I have trained myself to always pull the plug before I enter the furnace with tongs or any tool.

DESIGN
ALTERATIONS

4

A great many furnaces and kilns have been built in home shops with standard firebricks as the main construction material. They present themselves very well for nearly any type of fuel-burning furnace, but a problem arises when it becomes necessary to provide a groove for an electric heating element... It is likely that a considerable savings in cost and time can be had by using castable refractory material to mold the various parts of the furnace and assemble them with grout on a metal frame.

In figure 33 is a furnace of number 1 size with an extra section of body in place. This adds to the size of the heating chamber without increasing the electrical capacity. Of course, heating is slower, but very nearly

FIGURE 34

the same maximum temperature is finally reached. Such a unit is very good for some low-firing ceramics and for some types of glass work, and certainly fine for some forms of roasting and drying.

Two of my number 1 furnaces and a number 4 furnace are seen in figure 34. Standard clay graphite crucibles are seen on top of the lids. The center unit is seen with two heating sections and a deep metal pot that is made from a length of pipe with a disc welded to its bottom. The combination of two small units actually exceeds the number 4 furnace in melting capacity; its extra depth makes it a more versatile unit for other purposes as well.

Another practical variation is to rest the body on a pair of bricks with a metal grid to support an inverted investment mold. The furnace is run at a low heat until all of the wax has melted and run out. Then it can be reassembled in the normal mode to calcine the mold. (Books on ''lost-wax'' investment casting can provide more information about the calcining process.)

These are only a few of the possible alternatives to the basic design. Special needs will dictate other shapes, sizes and arrangements. There is hardly a limit to Li'l Bertha's possibilities. If the job is at all appropriate to do with an electric heating unit, Li'l Bertha can be designed to do it. An extra section can be made up to rest on the heating section, and it can be of any size or shape. You need only let your mind imagine the shape and size of the chamber, and you can easily make it to suit the job at hand.

MUFFLE FURNACES

One of the advantages of an electric furnace is that there are no products of combustion to corrupt the charge, so work that would usually require a muffle in a fuel-burning furnace can be done. It is possible that some component of the charge may attack the

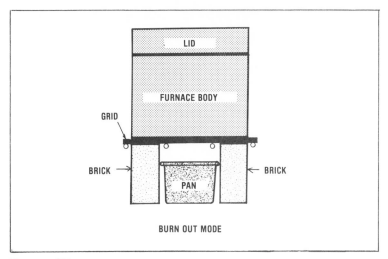

FIGURE 35

heating element, so a muffle may still be needed for some types of work.

The construction of a ceramic muffle requires considerable skill and knowledge, and I am not qualified to offer instruction in such work. Most likely a local potter can prepare a muffle to suit your needs at very moderate cost.

One process that may cause trouble is carburizing steel for the purpose of case hardening. This requires packing the steel in carbon granules, such as charcoal, charred leather or bone charcoal, and heating it to a red heat for a rather long time. Carbon will attack the heating elements and embrittle them, so a muffle is appropriate for such work. You can make a muffle of standard pipe with a cap on both ends. A hole is drilled in the center of one end to allow gas to escape and for insertion of test wires. It is obvious that a gas-tight muffle would be likely to explode when heated, so don't fail to provide the vent hole even if test wires are not used. If the vent hole in the furnace lid is left open, the carbon vapors that escape from the test hole in the muffle should leave the furnace without harming the element. A number of wires are set in the test hole so that they are held in the midst of the charge. A wire is pulled out from time to time to check its radiant color. When a bright red heat is attained, the charge is hot enough to accept some of the carbon in the pack. Of course, work of the same nature can be done with a ceramic muffle, and sizes and shapes can vary.

RECTANGULAR FURNACES AND KILNS

A great many furnaces and kilns have been built in home shops with standard firebricks as the main construction material. They present themselves very well for almost any type of fuel-burning furnace, but a problem arises when it becomes necessary to provide a groove for an electric heating element. The choice naturally runs to insulating brick for operating economy, and fortunately they are soft enough so that some sort of groove can be cut in them with ordinary hand tools. This is a bit tedious though, and insulating firebrick is quite costly. A considerable savings in cost and time can be had by using a castable refractory material to mold the various parts of the furnace and assemble them with grout on a metal frame. One such plan is seen in figure 36.

One method for molding the inside lining with grooves for the element is to prepare wooden forms to represent the shape of the desired groove. There is an advantage to an oversized groove with a dropped channel, so the forms can be made from common dowel stock and wooden strips as seen in figure 37. Note that the space between the round part and the square part is filled with a material like body putty or wood putty.

When the size of the furnace and the number of grooves has been determined, a simple wooden form can be made to cast each side. Notches are made in the edges so that the element groove forms can be located as desired in the face of the refractory block. These are made to slip in and out easily, so that they

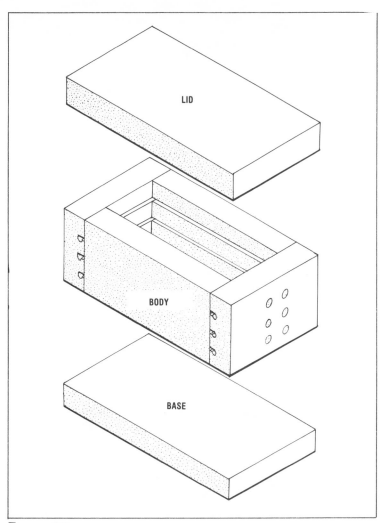

FIGURE 36

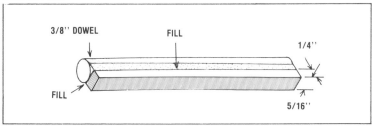

3/8'' DOWEL

FILL

FILL

1/4''

5/16''

FIGURE 37

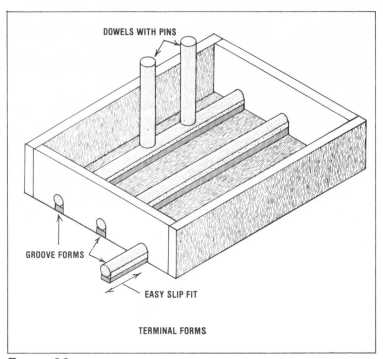

DOWELS WITH PINS

GROOVE FORMS

EASY SLIP FIT

TERMINAL FORMS

FIGURE 38

can be pulled out as soon as the refractory material has attained its initial firm set. It would be quite difficult to pull them out if they were left in during the initial curing period. A sample drawing of such a form is seen in figure 38. Note the dowel forms for the terminals, which should be on each row of forms.

It will be possible to work out many variations to the element plan, but a simple design would provide for varying heats by staging two or more elements as shown in figure 39. Of course, the infinite range control can be used on any element that is within the current-carrying capacity of the control. It is also possible to work out a stepped element installation with a very broad range of possible heats. Either separate switches or special function rotary switches can be used to provide a variety of steps or combination of elements in series or parallel. Of course the

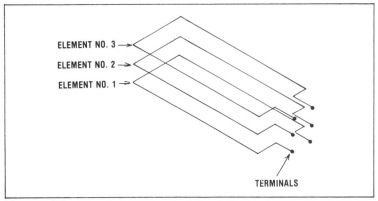

ELEMENT NO. 3
ELEMENT NO. 2
ELEMENT NO. 1

TERMINALS

FIGURE 39

same plan can be carried out with a large round furnace as well as a rectangular model.

When the inner lining and firing chamber have been worked out, it is a simple matter to devise a base of angle iron on which to rest the unit, and a sheet

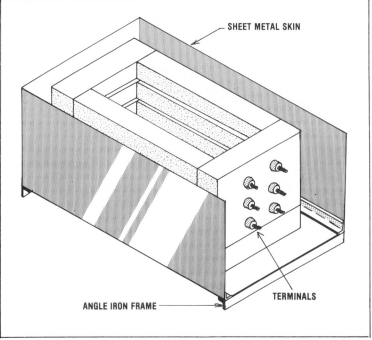

SHEET METAL SKIN

TERMINALS

ANGLE IRON FRAME

FIGURE 40

metal housing to support the controls. This housing can also be packed with an insulating material like vermiculite to increase the efficiency of the furnace, and to cut operating costs. A partial view of such a unit is shown in figure 40.

By using the castable refractory material, you can form the ready-to-use parts for the lining and heating chamber. A large part that must carry a load can be reinforced with wire mesh or metal rods so that thinner sections of stronger materials can be used where needed. The remainder of the unit can be of low-cost material of higher insulating value. Such a unit, as shown in figure 40, could be surrounded by vermiculate between the refractory lining and the sheet metal skin. The top surface can be sealed with a layer of castable refractory, and the lid might be supported on a counter-balanced form of angle iron. There is hardly a limit to size and shape when you build with castable refractories.

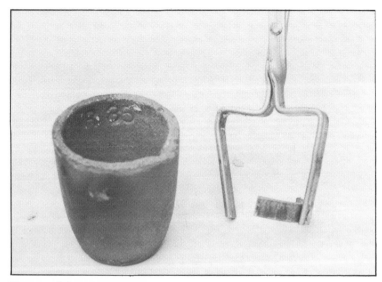

FIGURE 41

HANDLING
HOT STUFF

5

In almost every case of melting metal you will have a surplus to dispose of because you can't let it freeze in the pot. An ingot mold can be made of castable refractory. You can use a buscuit tin to cast handy-sized ingots. You can mold ingot-sized pockets in a sand bed to dispose of excess metal.

All of us have experienced painful burns to the skin at some time, but the severe damage caused by the concentration of heat found in this type of equipment deserves extra caution. *You simply must be prepared to handle the load safely and efficiently, if you are to avoid serious injury.* Properly made tongs, of sufficient weight and length, are one way of handling the load safely.

It would not be possible to offer detailed plans for building all types of tongs in this manual, but the crucible tongs shown in figure 41 can serve as a guide.

These tongs are shown with a number 1 clay graphite crucible which they were made to fit. The material used was a length of 5/8'' x 3/16'' flat steel stock, which is sold at building supply outlets for use in the end of chain link fence. The pivot is a 1/4'' iron rivet that is cold-set with an ordinary one-pound hammer. After the pivot joint was made, I clamped the tongs in a vise and gave it a quarter-twist so that their wide dimension would be against the side of the crucible. Then I bent the jaw straps to conform to the

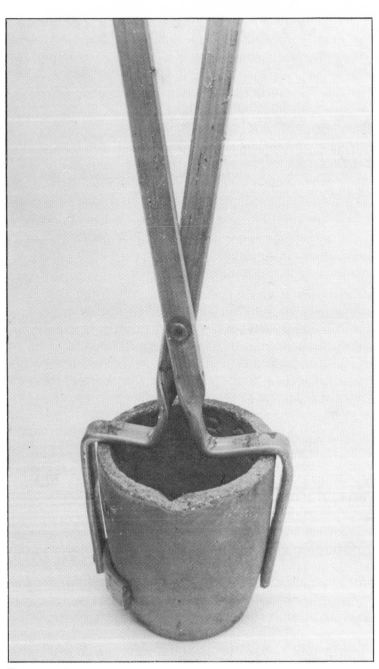

FIGURE 42

shape of the crucible and cut off the excess length. As a final step, I formed a short length of the same stock to fit the curve of the crucible, and then riveted it to one of the jaws to contact the pot just below the bilge. Note in figure 42 that the tongs touch the pot only below the bilge, and not at the rim. It is vital that you do not apply pressure to the rim of a ceramic crucible, because to do so may very well burst the pot and dump the molten charge of metal on your feet.

Of course larger and heavier tongs are required for larger pots. The light weight tongs shown here were made entire by cold forging; there was no need to heat the metal to shape it. Heavier tongs would require some real blacksmith work, but the electric furnace can be used as a forge, provided that you are extremely cautious about contacting the heating element while it is electrically energized. Some of the work on tongs can be done by welding, but cold-set rivets can be just as effective if carefully set in place.

If you use a metal pot, such as one welded up of steel pipe, you can use tongs that grip the rim of the pot because there is no danger of bursting the pot. Handles of reasonable length are needed because of the intensity of the heat radiated from the pot. Ordinary welders' gloves give the hands some protection, but the heat may penetrate the gloves if you must carry a pot an unusual distance with pliers or short tongs. *Always be prepared to set the pot down safely when the need arises.* A moment of panic can be tragedy. One or more bricks should be placed conveniently, so that the pot can be set down at any instant without any danger of tipping it.

In almost every case of melting metal, you will have a surplus to be disposed of because you can't let it freeze in the pot. An ingot mold can be made of castable refractory. You can use a biscuit tin to cast handy sized ingots. You can mold ingot-sized pockets in a sand bed to dispose of excess metal.

In spite of caution, there is always danger that a ceramic pot may break or leak before you can dispose of the charge. *The furnace must be set on a non-combustible surface, in the event that a charge leaks out of it.* You must realize the danger of molten metal running into cracks in a wooden floor, or running along a sloping floor to contact a combustible wall. It is very dangerous to throw water on molten metal, so you must be prepared with an amount of sand to cover a spill. The furnace should rest on a bed of sand if the floor in the shop is wooden, and the entire working area should be a bed of sand at least 2" thick. No molten metal should be carried beyond the sand bed.

The lid of the furnace will reach a temperature as high as the charge, and to set it on a combustible surface will surely start a fire. A pair of dry bricks are a convenient platform on which to set the lid, and it is even better if the bricks rest on a sand bed. I've found it handy to set the lid on a pair of bricks, and then the lid provides a handy surface on which to set the crucible, if it must be set down to get a better grip on it.

When the furnace has reached temperatures that are above 1000° F, it is not possible to tell when the element is electrically charged because it will glow at all times. I've trained myself to pull the plug so that there is no danger of contacting a live electrical element with the tongs. Even though the shock may not do serious damage in itself, a sudden jerk with a pot of molten metal could be very dangerous. An electrical interlock, that would automatically disconnect power when the lid is removed, would be an excellent safety control in the event that you are designing a more elaborate unit.

The furnace is especially handy for firing small ceramic articles because the infinite range control allows slow, gradual temperature rise. I've found that I must leave the vent hole open for the first two or three hours when firing a damp load. Once it reaches a red heat, you can cover the vent hole and run the temperature up as fast as you like.

It would be very difficult to place a limit on the possible uses of this type of unit. Of course, any corrosive material would damage the lining and element. More important are the fumes that may be liberated when a substance has been heated. *It will be very important to know as much as possible about the substance you heat, and to protect yourself from any potentially harmful fumes or vapors.*

It is certainly vital to make sure that no flammable or explosive vapors are in the work area. Many flammable vapors will ignite at temperatures even lower than those developed by this type of unit.

Surely not all dangers and possibilities have been covered, but it should be plain that you have here a useful tool that can open a new world of shop experience. It is my hope that you will improve Li'l Bertha's design to suit your own special use, and thay many of you will find ways to make new methods and processes possible in small home shop operations.